IMAGES
of America

WRIGHT BROTHERS
NATIONAL MEMORIAL

IMAGES
of America

WRIGHT BROTHERS
NATIONAL MEMORIAL

Douglas Stover and Darrell Collins

ARCADIA
PUBLISHING

Published by Arcadia Publishing
Charleston, South Carolina

Printed in the United States of America

Library of Congress Control Number: 2019940812

For all general information, please contact Arcadia Publishing:
Telephone 843-853-2070
Fax 843-853-0044
E-mail sales@arcadiapublishing.com
For customer service and orders:
Toll-Free 1-888-313-2665

Visit us on the Internet at www.arcadiapublishing.com

This book is dedicated to the memories of legendary volunteers
Joe Hardman, Bud Brown, Jerry Raveling, and Father Bill—
who nearly have 75 years of combined service to the park.

CONTENTS

ACKNOWLEDGMENTS

We are pleased to make available this collection of historical images of the Wright Brothers National Memorial, Kill Devil Hills, North Carolina, with the help of many sources and research material from the National Park Service Outer Banks Group and the Outer Banks History Center. Special thanks go to Jami Lanier, cultural resource manager, National Park Service Outer Banks Group, for allowing access to the Wright Brothers National Memorial Collection of the National Park Service Archives. Special thanks are given to Tama Creef and Stuart Parks, archivists at the Outer Banks History Center, a regional archives and research library in Manteo, North Carolina.

We want to thank Caroline Anderson, editor at Arcadia Publishing, and the production staff for allowing us to research and write this book and working with us through all stages of publication.

Unless otherwise noted, all images used in this book come from National Park Service, Wright Brothers National Memorial.

INTRODUCTION

Time flies. It is hard to imagine that 2019 marks the 50th anniversary of the first moon landing. It was an event labeled by historians and scholars as mankind's greatest achievement, second only to the first powered flight at Kitty Hawk. It happened at a time when America needed it the most; the country was divided by the war in Vietnam and slowly recovering from the assassination of Pres. John F. Kennedy. The moon landing was the most watched television event of the 20th century. Over 600 million people around the world witnessed Neil Armstrong step foot on the moon and will forever remember the following immortal words: "That's one small step for a man, one giant leap for mankind." The age-old dream of mankind came true that hot summer night 66 years after the world's first successful controlled, heavier-than-air powered flight.

Congress authorized the construction of the Kill Devil Hill Monument on March 2, 1927. It was transferred from the US War Department to the National Park Service on August 10, 1933, and was renamed the Wright Brothers National Memorial on December 4, 1953. The 428-acre national memorial site is located in Kill Devil Hills, North Carolina, Dare County, off US Route 158. Nearly 500,000 people a year visit the Wright Brothers National Memorial.

Two of seven siblings, Wilbur Wright was born in Millville, Indiana, on April 16, 1867, and Orville Wright was born in Dayton, Ohio, on August 19, 1871. Though they were the owners of a bicycle shop, the brothers had their sights on the sky and were determined to make technological advances in controlled powered flight. Inspired by famous glider Otto Lilienthal, who tragically died in a glider crash, the Wright brothers used gliders as a starting point.

To operate a glider, one needs regular breezes. With the help of the US Weather Bureau and the recommendation of American civil engineer and aviation enthusiast Octave Chanute, the brothers selected Kill Devil Hills, North Carolina, in the Outer Banks. The area met their list of criteria; it was an isolated, vegetation-free space with sand dunes and steady wind. Located four miles south of Kitty Hawk (the nearest settlement), Kill Devil Hills was the perfect location for the Wright brothers to perform their flying experiments. Though the brothers preferred experimenting in a nonpopulated area, it did make it more challenging to get supplies and thus were forced to return home and regroup at the end of every flying season. So from 1900 to 1903, Wilbur and Orville Wright split their time between their home in Dayton, Ohio, and their temporary camp in Kill Devil Hills.

The Wrights first arrived in Kitty Hawk, North Carolina, in the fall of 1900. On that brief trip, they tested only about 12 free glides, with Wilbur lying flat on the lower wing. They used Big Kill Devil Hill to facilitate the glider's takeoff. The Wright brothers were unsatisfied with the lift of the glider, so they constructed a larger one for their experiments in 1901.

In the summer of 1901, the larger glider had improved in some areas, like landing, but not all. After numerous tests, the lift generated by the wings was only a third of what they predicted using Lilienthal's tables (coefficients of lift). Once back home, the brothers built a wind tunnel, which allowed them to test different shapes and airfoil curves of wings.

The 1902 glider featured a fixed, rear vertical rudder and improved wing design that gave them a vertical lift; however, there were still problems with lateral (roll) control, also known as "wing warping." According to the National Park Service, between 1900 and 1902, "the Wrights glided off [Kill Devil Hills] more than 1,000 times." At home in Ohio, the Wrights conducted experiments with propellers and began to build their 1903 flyer, which had a water-cooled vertical four-cylinder engine and a lightweight wooden glider for its body.

On September 25, 1903, Wilbur and Orville arrived at Kitty Hawk. There, the brothers assembled their 1903 machine's engine and started their experiments. Unfortunately, in November, the propeller shafts broke twice, forcing the brothers to return to Dayton to repair parts. After making the necessary repairs and traveling back to North Carolina, on December 14, 1903, because of the lack of wind, the brothers tried launching the flyer off the side of Big Kill Devil Hill. They tossed a coin to determine who would make the first trial, and Wilbur won. What occurred that day was the first stall of a powered airplane; the flight only lasted 3.5 seconds and covered 105 feet. On landing, the undercarriage was damaged. It took three days to repair the damage.

On Thursday, December 17, 1903, at 10:35 a.m., with Orville Wright at the controls of the flyer, the world changed forever. The first successful controlled, heavier-than-air powered flight lasted 12 seconds and covered 120 feet. In addition to the brothers, five other people witnessed the momentous occasion. US Lifesaving Service crew members Adam Etheridge, Will Dough, and John T. Daniels, who took the iconic "first flight" photograph using Orville's camera, were there. Also in attendance were area businessman W.C. Brinkley and local teenager Johnny Moore. After the first flight, the brothers made three additional trials. At 11:00 a.m., Wilbur made the second flight, 175 feet in 12 seconds. At 11:40 a.m., Orville made the third flight, 200 feet in 15 seconds. All the flights that morning reached a height of 12 to 14 feet off the ground. The fourth and last flight on December 17, 1903, was made by Wilbur Wright. He covered a distance of 852 feet in 59 seconds. With these successful flights, the Wrights not only changed the course of Kill Devil Hills and Kitty Hawk forever but also the world.

In the 1920s, after years of deliberation, the government decided to recognize Kill Devil Hills as the birthplace of aviation and honor the Wright brothers with a monument. In 1927, Pres. Calvin Coolidge signed a bill that assigned $50,000 to the construction of Kill Devil Hill Monument. The first formal celebration honoring the Wright brothers and their achievements was held on December 17, 1928, the 25th anniversary of the world's first powered flight. Over 3,000 people attended the ceremony, including representatives from 40 nations. Two events were scheduled for the ceremony, the laying of the cornerstone of the future national monument atop of Big Kill Devil Hill and the unveiling of a six-foot-high granite boulder marking the spot where the Wrights made the first flight on December 17, 1903.

The US Commission of Fine Arts had review authority over all construction within the federal government and held a competition for the design of the monument. Thirty-six entries were submitted, and in February 1930, New York architects Robert P. Rogers and Alfred Easton Poor won the competition and the $10,000 prize. Construction began 18 months later, and by then, its budget increased by an additional $150,000. In the end, the 1,200-ton granite monument cost a total of $213,000 and was completed by November 1932. Its dedication was held on a windy November 14, 1932, with Orville Wright as the guest of honor; sadly, Wilbur had passed away in 1912.

Later, in 1953, the National Park Service reconstructed two of their gliders to celebrate the 50th anniversary of the Wright brothers' first flight, and from this original idea, a visitor center was constructed at the site of the Wright Brothers National Memorial. The visitor center is of modern design and was built as part of the National Park Service's Mission 66, a modernization and expansion program. The national memorial was placed in the National Register of Historic Places on October 15, 1966, and the visitor center, designed by Ehrman Mitchell and Romaldo Giurgola, was designated a National Historic Landmark on January 3, 2001, for its architecture and its importance to the National Park Service's Mission 66 program.

In 2003, the Wright Brothers National Memorial hosted the six-day First Flight Centennial Celebration from December 12 to December 17. In attendance as guests of honor were the following: flight enthusiast John Travolta, Pres. George W. Bush, Neil Armstrong, Buzz Aldrin, and Chuck Yeager. For the celebration, the National Park Service constructed the 20,000-square-foot First Flight Centennial Pavilion; it had three buildings and was constructed at a cost of $2 million. At the pavilion, visitors could explore exhibits sponsored by NASA, the Federal Aviation Administration, and the Experimental Aircraft Association. The State of North Carolina donated a bronze sculpture to the national memorial, and it was dedicated during the centennial celebration. The life-sized sculpture, created by Stephen H. Smith, depicts the *Wright Flyer*, the brothers, and the five other people who witnessed the first successful flight on December 17, 1903.

Today, most guests to the Wright Brothers National Memorial start their trip at the Wright Brothers Visitor Center to review the different exhibits and artifacts and see the full-scale 1903 flyer replica. From the visitor center's patio, guests can explore the outdoor space independently or opt to join a 30-minute ranger-led walking tour. On the walking tour, visitors learn about what life was like for the brothers while stopped at the reconstructed 1903 living quarters and hangar and will hear of their successes while stationed in front of the First Flight Boulder. Once the talk concludes, visitors are encouraged to traverse the walkway up to the Wright Brothers National Memorial. There, one will enjoy breathtaking views of not only the memorial, but also the surrounding area and the First Flight Airstrip. Most guests end their visit on the opposite side of Big Kill Devil Hill at the life-size sculpture that recreates the historic 1903 flight. There, they can take their own "first flight" photograph with the Wright brothers.

One

WRIGHT BROTHERS AT KITTY HAWK

Wilbur and Orville Wright are the inventors of the airplane. Two of seven siblings, Wilbur was born in Millville, Indiana, on April 16, 1867, and Orville was born in Dayton, Ohio, on August 19, 1871. They were the sons of Milton Wright, a circuit preacher who later became a bishop in the United Brethren Church of Christ, and Susan Koerner Wright, a homemaker. The brothers inherited their mechanical ingenuity from their mother. In a letter to American civil engineer and aviation enthusiast Octave Chanute, dated May 13, 1900, Wilbur Wright wrote, "For some years I have been afflicted with the belief that flight is possible to man. My disease has increased in severity and I feel that it will soon cost me an increased amount of money if not my life." Three years later on December 17, 1903, the Wright brothers made the world's first successful controlled, heavier-than-air powered flight. For centuries, humankind had dreamed of flying. Within the lifetime of a generation, those dreams came true. (Courtesy of the Library of Congress.)

On November 26, 1899, Wilbur wrote the US Weather Bureau in Washington, DC, inquiring about windy areas in the country to conduct flying experiments. On December 4, the chief of the Weather Bureau sent data on windy spots in the country. The number one windiest spot on the list was the city of Chicago and number six was Kitty Hawk, North Carolina. (Courtesy of the Library of Congress.)

On August 3, 1900, Wilbur wrote a letter to the Weather Station at Kitty Hawk. J.J. Dosher, the manager of the station, responded to the letter about the winds and the topography of the land. The letter was then passed on to Kitty Hawk postmaster William Tate, who, in another letter, extends the Wrights a personal invitation to come to Kitty Hawk to conduct their flying experiments. He also describes the area as isolated with very few trees along with high dunes. (Courtesy of the Library of Congress.)

ORVILLE AND WILBUR WRIGHT WERE TRANSPORTED TO KITTY HAWK-1903 BY HATTIE CREF

At the turn of the 20th century, the only way to get to Kitty Hawk was by sailboat. After a seven-day journey, Wilbur arrived at Kitty Hawk on September 13, 1900. Fifteen days later, Orville arrived and remarked when he first sees it, "This is how I always imaged the Sahara Desert looked."

"This is a great country for fishing and hunting," Orville wrote to his sister. "The fish are so thick you see dozens of them whenever you look down into the water. The woods are filled with wild game, they say; even a few 'b'ars' are prowling about the woods not far away." (Courtesy of the Library of Congress.)

Postmaster William Tate; his wife, Amanda (also known as Addie); their daughters Irene and Pauline, aged three and two, respectively; and an unidentified woman (possibly Maxine Cogswell, Addie's sister) are pictured on the porch of the Kitty Hawk Post Office in September 1900. On the morning of September 13, 1900, there was knocking at the door of William Tate's house in Kitty Hawk, North Carolina. When Tate appeared, Wilbur took off his cap and introduced himself as Wilbur Wright of Dayton, Ohio. (Courtesy of Wright State University Libraries.)

The brothers set up camp in a tent a half a mile south of the Tate house. In early October 1900, they began their experiments flying their glider both as a kite and a man-carrying glider. Just before leaving Kitty Hawk, the brothers spent an entire day gliding from the slopes of Big Kill Devil Hill, making about a dozen manned flights. That day, total time in the air was two minutes. Overall, they were very pleased with their experiments and looked forward to returning the next year. (Courtesy of the Library of Congress.)

14

Two

THE WRIGHT BROTHERS BEGIN THEIR EXPERIMENTS (1900–1911)

In early October 1900, the Wright brothers began their flight experiments, setting in motion a series of events that allowed the first generation of flight to travel from Kitty Hawk to the moon in the lifetime of a human being. A year earlier, Orville Wright explained in detail their first interest in the subject of aeronautics. He said, "Our first interest began when we were children. Father brought home to us a small toy actuated by a rubber spring which would lift itself into the air. We built a number of copies of this toy, which flew successfully . . . but when we undertook to build the toy on a much larger scale it failed to work so well." This was the seed planted in their young minds by their father that, after years of successes and failures, would cultivate into one of the greatest inventions in the history of world. (Courtesy of the Library of Congress.)

The Wrights' 1900 glider was covered with imported French sateen, a high-quality cotton fabric with a silky look. Local ladies who lived in Kitty Hawk had never seen such fine cloth. When the Wrights departed Kitty Hawk in late October, Addie Tate salvaged the cloth, gave it a good washing, and made two dresses for her young daughters. (Courtesy of the Library of Congress.)

In the winter of 1901, the brothers built a six-foot wooden wind tunnel with a glass window on top to observe the movement of the balances. A two-bladed belt-driven fan was powered by a gas motor that created a wind speed between 25 and 35 miles per hour. The data compiled from the wind tunnel would lead to the success of their 1902 glider, the world's first true flying machine, and their 1903 flyer. (Courtesy of the Library of Congress.)

The 1901 glider experiments were very disappointing. The Wrights had built the largest glider anyone had ever tried to fly. However, the lift generated by the wings was only a third of what they predicted using famous glider Otto Lilienthal's tables (coefficients of lift), and there were unforeseen problems with lateral (roll) control, also known as "wing warping." When they left Kitty Hawk on the return train ride back to Dayton, Wilbur said to Orville, "Not within a thousand years would man ever fly." Later, Wilbur wrote about the 1901 season, "We consider our experiments a failure." (Courtesy of the Library of Congress.)

Pictured on October 24, 1902, Wilbur Wright recovers from making a right turn in their 1902 glider. They successfully flew the glider 1,000 times and set a new gliding world record of traveling 622.5 feet in 26 seconds. The 1902 glider was the first machine in recorded history to make a coordinated turn, recover from the turn, and fly at right angles to the wind. (Courtesy of the Library of Congress.)

The brothers returned to Kitty Hawk on August 28, 1902, to find their old building from the year before had sank two feet in the sand on each end. Over the next several days, they raised the building and began construction of a new addition. Once the addition was completed, the brothers began assembling their 1902 glider. (Courtesy of the Library of Congress.)

The kitchen of their 1902 camp building was well stocked with canned goods since fresh meat and vegetables were not readily available. Daily, the brothers shared the chores of cooking and cleaning. (Courtesy of the Library of Congress.)

Wilbur was the better cook of the two brothers. In the foreground is the last-known photograph of the Wright brothers' 1901 glider before parts were used in their 1902 glider. (Courtesy of the Library of Congress.)

The 1903 flyer was powered by a 12-horsepower aluminum-block gasoline engine. To calculate the air speed of the flights, there were three instruments on the machine: a revolutions-per-minute counter, which was connected to the main drive shaft to determine the efficiency of the propellers; a stopwatch; and a Richard anemometer for time and distance. (Courtesy of the Library of Congress.)

The Wrights returned to Kitty Hawk in late September 1903 after months of delays. The flyer was ready for flight in the late afternoon on December 12, 1903. (Courtesy of the Library of Congress.)

On September 26, 1903, the brothers began construction of a new building to house their 1903 flyer. The building was completed on October 5, 1903, and the brothers commenced assembling their 1903 machine. (Courtesy of the Library of Congress.)

On December 14, 1903, because of the lack of wind, the brothers tried launching the flyer off the side of Big Kill Devil Hill. They tossed a coin to determine who would make the first trial, and Wilbur won. What occurred that day was the first stall of a powered airplane; the flight only lasted 3.5 seconds and covered 105 feet. On landing, the undercarriage was damaged. It took three days to repair the damage. (Courtesy of the Library of Congress.)

On Thursday, December 17, 1903, at 10:35 a.m., with Orville Wright at the controls of the flyer, the world changed forever. Instead of having the flyer take off from Big Kill Devil Hill, the brothers launched it on a short rail, which was laid on level ground. The first flight lasted 12 seconds and covered 120 feet. The photograph was taken by one of the witnesses, John T. Daniels, a member of the US Lifesaving Service, which is now known as the US Coast Guard. Orville set the camera up, and all John had to do was to squeeze the bulb. The brothers did not develop the glass plate until they returned home to Dayton; there, in the darkroom, they realized John captured the very first seconds of true powered flight. That moment, nicknamed the "Miracle at Kitty Hawk," was preserved for generations to come. (Courtesy of the Library of Congress.)

After the first flight, the brothers made three additional trials. At 11:00 a.m., Wilbur made the second flight, 175 feet in 12 seconds. At 11:40 a.m., Orville made the third flight, 200 feet in 15 seconds. All the flights that morning reached a height of 12 to 14 feet off the ground. (Courtesy of the Library of Congress.)

The fourth and last flight on December 17, 1903, was made by Wilbur Wright. In Orville's diary, he recounts the flight as follows: "The machine started off with its ups and downs as it had before but by the time he had gone three or four hundred feet he had it under better control . . . it proceeded in this manner out about 800 feet . . . when it began its pitching again and suddenly darted into the ground . . . the front frame was broke . . . distance covered 852 feet in 59 seconds." (Courtesy of the Library of Congress.)

RECEIVED at   *17*

76 C KA CS 33 Paid.   Via Norfolk   Va

itty Hawk N C Dec 17

ishop M Wright

       7 Hawthorne St

uccess four flights thursday morning all against twenty one mile

ind started from Level with engine power alone average speed

hrough air thirty one miles longest 57 seconds inform Press

ome home Christmas .   Orevelle Wright   525P

More flights were planned that day, but at the end of the fourth flight, the machine was struck by a gust of wind, destroying it beyond repair. After dinner, the brothers walked down to Kitty Hawk and sent their father, Bishop Milton Wright, the famous telegram announcing the success that day. (Courtesy of the Library of Congress.)

By February–March 1908, the Wrights had signed contracts with the US War Department and French capitalists. Having not flown since October 1905, the brothers returned to Kitty Hawk for the last time together to do some practice flying in 1908. They modified their 1905 flyer to carry a passenger. On May 14, 1908, Wilbur Wright makes the world's first passenger flight, carrying their mechanic Charles W. Furnas. The flight was Furnas's first airplane ride. On that same day, Orville with Furnas made a 2.5-mile flight. (Courtesy of the Smithsonian Institution.)

The flights made at Kitty Hawk in 1908 were witnessed by newspaper reporters from the *New York Herald, New York American,* and *London Daily Mail*; a photographer from *Collier's Weekly*; locals; and men from the Kill Devil Hills Lifesaving Station. People around the world read newspaper accounts of the Wright brothers' flights at Kitty Hawk, North Carolina, in 1908. (Courtesy of the Library of Congress.)

In October 1911, Orville Wright brought their last flying machine to Kitty Hawk. He had been working on an automatic stabilizer (today's automatic pilot), and his test bed was a glider with a wingspan of 32 feet with a wing area of 300 square feet and weight of 170 pounds. Because newspaper reporters visited the camp daily, it was never tested in the glider. Patent No. 1,075,533 was granted on October 14, 1913, on "Device for maintaining automatic stability." (Courtesy of the Library of Congress.)

Between October 16 and October 26, 1911, Orville made 90 glider flights off the Big Kill Devil Hill, West Hill, and Little Hills. On October 24, 1911, Orville set a new world soaring record at 9 minutes and 45 seconds in a 50-mile-per-hour wind off Big Kill Hill. That record would stand until 1921, when it was surpassed in Germany. (Courtesy of the Library of Congress.)

When Orville Wright returned to Kitty Hawk with the brothers' last flying machine, he brought along with him his older brother Lorin Wright, his son Horace "Buster" Wright (pictured above), and his English friend Alec Ogilvie. (Courtesy of the Smithsonian Institution.)

Three

MONUMENT TO AN IMPOSSIBLE DREAM

SECTION

NORTH ELEVATION OF MONUMENT

Serious consideration to memorialize the Wright brothers and the first flight at Kitty Hawk started in the mid-1920s. World War I had just ended, and the airplane had proved to be a formidable instrument of change on the battlefield. After the war, a surplus of airplanes cheaply priced gave rise to a new breed of aviators—barnstormers—who would thrill crowds of spectators across the nation and give most Americans their first glimpse of an airplane. (Courtesy of the National Archives.)

Spare vegetation surrounds Big Kill Devil Hill in this photograph. To the left is the tree line of a maritime forest taking hold on the back side of the barrier island. Located along the Outer Banks of North Carolina on the back side of the island, away from the wind and salt spray, the primary trees that grow in the maritime forest are pine, cedar, and live oak.

This rare photograph shows the four sand dunes known as the Kill Devil Hills of Kitty Hawk (Big Kill Devil Hill, West Hill, and Little Hills). Between 1900 and 1903, the Wright brothers made over 1,000 glider flights off the hills, setting all the world records in gliding at the turn of the 20th century.

28

In the distance is Big Kill Devil Hill. It is part of the common geological feature that occurs along the Outer Banks called *médanos*, which is derived from the Spanish word for "coastal sand hills." This unvegetated migratory sand dune has tremendous amounts of sand shaped by the wind and can reach heights several hundred feet tall. Four miles south is North Carolina's most famous *médanos*, Jockey's Ridge State Park, the largest sand dune on the East Coast.

In 1926, the property encompassing Big Kill Devil Hill was owned by Frank Stick (pictured) & Associates, a real estate development company from Asbury Park, New Jersey. Stick and his associates donated 200 acres of land to the government for the purpose of establishing the national monument. In 1927, an acre of land sold for $10. By the time the monument was dedicated in 1932, the price per acre had risen to $40 per acre.

Here, the Kill Devil Hill Monument's cornerstone is in place on Big Kill Devil Hill in preparation of the 25th anniversary celebration. The cornerstone was unveiled during the celebration and was later relocated. A recent X-ray search for the cornerstone revealed its location on the northeast point of the monument, facing the First Flight Boulder.

Before the construction of the monument could begin, Big Kill Devil Hill had to be stabilized to stop its migration toward the southwest. Stabilization of the hill began in January 1929. Captain Kindervater, of the US Army Quartermaster Corps, started by plowing and putting down pine straw, rotten leaves, and wood mold. He then planted Bermuda grass seed, bitter tannic, yaupon, myrtle, pine, live oak, and sumac; covered it with brush; and fertilized the area heavily. The stabilization of the 26-acre hill was finished two years later in February 1931.

In 1928, when this photograph was taken, Big Kill Devil Hill was still a live sand dune. The prevailing winds on the Outer Banks blew from the northeast, causing the hill to migrate toward the southwest. Before the monument could be constructed atop the hill, the dune would migrate 450 feet to the southwest in 25 years.

A grassroots effort on the local and state levels started in the mid-1920s to build a monument to the Wright brothers. More than anyone else, the newly elected US representative Lindsay Carter Warren, a native of Washington, North Carolina, envisioned the establishment of a national monument at Kill Devil Hill. On December 17, 1926, Representative Warren introduced his bill to the House, and on the same day, Sen. Hiram Bingham, former governor of Connecticut, introduced a bill very similar to Warren's in the Senate.

On March 2, 1927, Pres. Calvin Coolidge signed the bill into law. Construction on the monument began in October 1931 and was finished in November 1932.

Over 1,200 tons of granite from the largest open-faced granite quarry in the world, located in Mount Airy, North Carolina, was used in the construction of the monument. The monument cost $285,000 to build.

The monument was dedicated to the Wright brothers on November 19, 1932, at the height of the Great Depression. Orville Wright, the surviving brother, was the guest of honor. It was the largest monument in America to be dedicated to a living human being. Wilbur Wright died in 1912, twenty years before the dedication. The monument that sets atop of Big Kill Devil Hill is a testament to the questing nature of the human spirit, as what the Wright brothers did here, changed the world forever. The inscription around the monument reads, "In commemoration of the conquest of the air by the brothers Wilbur and Orville Wright, conceived by genius achieved by dauntless resolution and unconquerable faith."

The Kill Devil Hill Monument was dedicated on November 19, 1932. Over 20,000 people were expected, but wind and rain prevented them from attending the dedication. Only about 1,000 people showed up. Standing in the back near the corner of the monument and wearing a dark hat was Orville Wright, the guest of honor. Standing to his right is Ruth Nichols, famous aviatrix, and to his left are his sister-in-law Ivonette Wright and her husband, Lorin Wright. Standing in front, with his hat in hands, is Lorin's son Milton Wright with his wife and sons, Wilkerson Wright and Milton Wright, Orville's grandnephews.

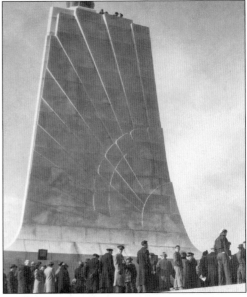

The white granite monument sets atop the 90-foot-tall Big Kill Devil Hill. Visitors to the site could climb the 60-foot monument to the observation platform on top. From 1983 to 1993, the monument was open every summer for visitors to climb. Once inside the first floor, there are two inscriptions on the east and west walls. Carved in the west wall, the inscription reads: "From a point near the base of this hill Wilbur and Orville Wright launched the first flight of a power-driven airplane on December 17, 1903."

The doors of the monument are stainless steel over nickel. Each of the eight bronze panels represent mankind's early attempts to fly and the ultimate conquest of air. The doors are part of the original design of the monument and cost $3,000 to fabricate in 1928. (Courtesy of National Geographic Society.)

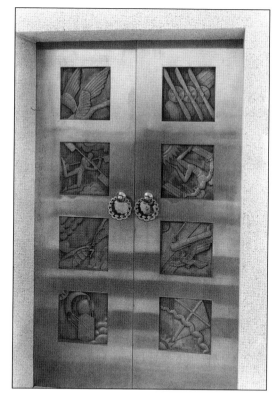

Two of the most striking features of the monument are the busts of Wilbur and Orville Wright, situated on the steps near the entrance of the structure. Sculptor Oskar Hansen, who is best known for his sculptures on and around the Hoover Dam, created the busts. The busts at the monument today are reproductions. The original busts were stolen in the late 1980s; however, they were mysteriously returned years later and are now housed in the National Park Service Archives.

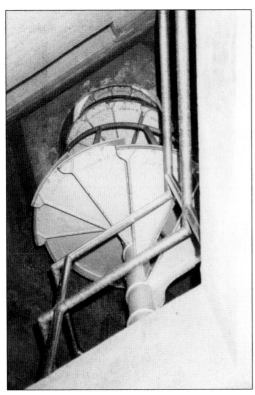

Between 1932 and 1960, before the visitor center was opened to the public, the monument served as the gathering place for visitors. The ranger on duty would provide history and orientation talks to the public inside the base of the monument. Visitors were encouraged to ascend the monument's interior stairs made of granite and steel to the small observation platform underneath the rotating beacon that comes on every night at dust.

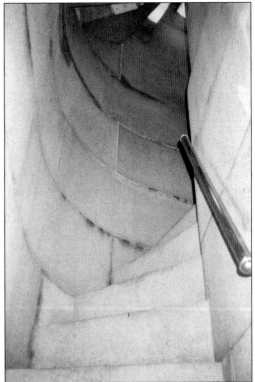

Four

COMMEMORATING
THE WRIGHTS

The earliest formal celebration honoring the Wright brothers and the world's first powered flight was held on December 17, 1928, the 25th anniversary of that original flight. Over 3,000 people attended the ceremony, including representatives from 40 nations. Two events were scheduled for the ceremony, the laying of the cornerstone of the national monument atop of Big Kill Devil Hill and the unveiling of a six-foot-high granite boulder marking the spot where the Wright brothers made the first flight on December 17, 1903.

After the laying of the cornerstone ceremony, the crowd moved down the hill to the site where the National Aeronautic Association had erected a six-ton granite boulder that marked the spot of the first airplane flight in world history. At the moment that Orville pulled the silk parachute to unveil the boulder, homing pigeons were released into the air and circled the crowd before heading back to their roost.

Seen here is William Tate, the original host to the Wright brothers at Kitty Hawk in 1900, addressing the crowd during the 25th anniversary celebration. Standing to the right of him are Orville Wright and Amelia Earhart. Tate, alone, persuaded the Wrights to choose Kitty Hawk for their experiments when he wrote, "I will take pleasure in doing all I can for your convenience & success & pleasure, & I assure you, you will find a hospitable people when you come among us."

INTERNATIONAL CIVIL AERONAUTICS CONFERENCE
WASHINGTON. D. C.
DECEMBER 12 – 13 – 14 – 1928

As part of the 25th anniversary celebration, an International Civil Aeronautics Conference was held in Washington, DC, December 12–14, 1928. The souvenir booklet from the conference was signed by Amelia Earhart, Orville Wright, and Clarence Chamberlin who piloted the first transatlantic flight of a passenger 14 days after Lindbergh's famous solo flight across the Atlantic.

Pictured is the cornerstone of the Kill Devil Hill Monument. The laying of the cornerstone was the first ceremony held during the 25th anniversary celebration. The cornerstone was later relocated once the monument was constructed, and the time capsule placed inside the stone was scheduled to be opened during the 100th anniversary celebration. (Courtesy of the US Department of Agriculture.)

Open-range grazing was a common practice in the Outer Banks in the late 1920s and early 1930s. Once plantings to stabilize the hill commenced, fencing was erected around the site to keep roaming livestock from eating the young vegetation. The state legislature passed the Livestock Act of 1934, which stopped open-range grazing in the Outer Banks. (Courtesy of the US Department of Agriculture.)

A crowd of nearly 3,000 people made its way up Big Kill Devil Hill to witness the laying of the cornerstone for the national monument. Assistant Secretary of War F. Trubee Davison, along with three of the witnesses to the first flight, John T. Daniels, Adam D. Etheridge, and Willie S. Dough, presided over the ceremony.

Five

THE PARK IN THE 1930s

This 1932 aerial view of the Kill Devil Hill Monument shows construction of the entrance road to the park off Colington Road. The road goes around the east side of the hill to a roundabout that leads northeast to the site of the first flight. The monument was constructed during the height of the Great Depression at a cost of $285,000. Funding ($90,500) for the park's infrastructure of roads, trails, buildings, and plantings was allocated by the Public Works Administration, created by Pres. Franklin D. Roosevelt, as part of the New Deal.

The US Commission of Fine Arts, established in 1910, had review authority over all construction within the federal government and held a competition for the design of the monument. Thirty-six entries were submitted for the design of the monument. New York architects Robert P. Rogers and Alfred Easton Poor won the competition and were awarded the $10,000 prize.

Built in the 1936 to accommodate more and more visitors to the monument, the original comfort station/bathroom was located on the south side of the hill. The structure was destroyed by fire in 1967.

The original park entrance gate and contact station, constructed in 1932, was located off Colington Road on the south side of today's park boundary. The only remains of the entrance are four concrete pillars that can be seen while riding down Colington Road.

By the late 1940s and early 1950s, the X-planes were breaking the speed of sound, and humankind was on the verge of touching the edge of space. The 50th anniversary of the world's first flight was quickly approaching, and more attention was placed on the two men whose vision and dream had changed the world. On December 1, 1953, just 16 days before the golden anniversary of the first flight, the name changed to the Wright Brothers National Memorial.

Pictured here are two legends of the Outer Banks; in uniform is Horace Dough, the first superintendent of the memorial who started as the caretaker in 1933 and spent the next 29 years at the monument. To his left with the camera is writer, publicist, and photographer Aycock Brown, the first director of the Outer Banks Tourism Bureau who took over 100,000 photographs that would bring the Outer Banks into the national spotlight and inspire tourism to the area in the 1950s and 1960s.

From the beginning, vandalism and trespassing after hours were problems at the site. The need for guards became apparent once incidents started to occur. In 1935, a budget of $13,500 was allocated out of funds from the Public Works Administration to build a superintendent quarters to provide better protection for the monument. During the First Flight Centennial Celebration in 2003, the building was used as the headquarters of the First Flight Commission and the First Flight Foundation for the planning of the event. The building was removed in 2013.

The original maintenance shop constructed in 1936 is still located along Colington Road just west of the original 1930 entrance to the park. Over the years, new additions have been added to the structure as well as fencing around the perimeter.

Funding from the Public Works Administration was also used to construct trails and roads around the hill. A total of $38,000 was budgeted to construct four symmetrical curving trails up to the monument. A .09-mile road around the base of the hill along with a parking lot on the south side provided access to the monument.

During World War II, US Navy blimps stationed out of Elizabeth City, North Carolina, were often seen flying over the memorial. Navy blimps played a critical role in coastal defense during the war. The blimps escorted convoys off the coast and carried sensors and weapons needed to detect and destroy German submarines that were stationed along the shipping lanes off the coast.

During the 44th anniversary celebration, Lois Smith, the granddaughter of John T. Daniels, one of the witnesses to the Wright brothers' first flight, participated in the wreath-laying ceremony. She later became a board member of the First Flight Society and spearheaded the wreath-laying ceremony every December 17 until she passed away in 2013. The second person to her right is Jimmy Doolittle, Medal of Honor recipient, who led the first bombing raid over Japan four months after the surprise attack on Pearl Harbor on December 7, 1941.

On December 17, 1947, the 44th anniversary celebration of the world's first powered flight was held at the monument. The man standing underneath Orville's name is John T. Daniels, one of the five men who witnessed the first flight of the Wright brothers and a photographer. Ironically, the man who took the famous photograph of the first flight that December morning died on January 31, 1948—24 hours after Orville died on January 30, 1948.

Orville Wright, the first man to fly, out lived his brother Wilbur Wright by 36 years. Orville lived through both world wars and saw the rocketry of World War II and the propeller give way to the jet engine. A few months before Orville died in January 1948, Chuck Yeager broke the sound barrier in the Bell X-1 in October 1947. In 1929, Orville wrote, "Aviation has gone beyond my dreams."

Six

THE MEMORIAL
(1950–1960)

On December 17, 1953, at Kitty Hawk, North Carolina, the world celebrated the 50th anniversary of the first controlled powered flight. The four-day celebration (December 14–17) brought national and international recognition to the achievements of the Wright brothers' gift of flight to the world. Over 15,000 people attended the celebration, including pilots; aviation pioneers; and state, national, and international dignitaries. Over 200 aircraft participated in an aerial display that highlighted the advancements of aviation over the past 50 years. Among the dignitaries were nieces and nephews of the Wright brothers; Gen. Frank Lahm, America's first military pilot who was trained by Wilbur Wright at College Park, Maryland, in 1909; and Jimmy Doolittle, Medal of Honor recipient who led the first bombing raid over Japan in April 1942.

PROGRAM

Golden Anniversary Observance

OF

MAN'S FIRST SUCCESSFUL POWERED FLIGHT

BY

ORVILLE and WILBUR WRIGHT

December 17, 1903—Kill Devil Hills, North Carolina

1903   1953

Sponsored by

KILL DEVIL HILLS MEMORIAL SOCIETY

AIR FORCE ASSOCIATION

NATIONAL PARK SERVICE

NORTH CAROLINA 50th ANNIVERSARY COMMISSION

December 14, 15, 16 and 17, 1953

KILL DEVIL HILLS

and

KITTY HAWK, NORTH CAROLINA

A four-day celebration was planned for the 50th anniversary. Activities for the event were sponsored by the National Park Service, Kill Devil Hill Memorial Society (now the First Flight Society), Air Force Association, and the North Carolina 50th Anniversary Commission.

Over 15,000 people attended the four-day 50th anniversary celebration. The celebration brought national and international recognition to the Wright brothers' first flights at Kitty Hawk. Over 200 aircraft participated at the event, highlighting the progression of modern aviation over the past 50 years.

Many aviation pioneers attended the 50th anniversary celebration. Among them was Igor Sikorsky, the Russian-born American who invented the first modern helicopter. In 1923, he founded the Sikorsky Aircraft Corporation, which was one of the first companies to sell helicopters for military and civilian operations.

This is an actual photograph taken during the 50th anniversary celebration. Sponsored by the Philips Petroleum Company, the reenactment of the Wright brothers' four flights was a popular event. Billy Parker, aviation pioneer, flew his 1912 home-built pusher biplane over the First Flight Markers. This remarkable photograph was taken as a B-47 Stratojet made a low pass, overtaking Parker's biplane.

In 1953, the entrance to the park was still located off Colington Road on the south side of the park. The historic landscape of the park was well established with trails, roads, and the commemorative open space between the monument north and the well-kept first flight grounds.

Before digging the foundations for the replicas of Wright brothers' living quarters and hangar, the site had to be thoroughly inspected for any historic artifacts. J.C. Harrington, archaeologist for the National Park Service, used a metal detector to search for metal objects that may have been left behind by the Wright brothers. From left to right are J.C. Harrington, unidentified, A.W. Drinkwater, and Supt. Horace Dough.

The drive up to the first flight area was paved with a roundabout circling the six-ton granite boulder. The boulder was elevated on a small mound and faced the monument up until the late 1960s.

The Wright brothers left no blueprints for their hangar and living quarters, but in letters home, they did describe them, the dimensions, and details of their construction method. The replicas were dedicated on December 14, 1953, Pioneers and Flyer Day, during the 50th anniversary celebration.

The Wright brothers' living quarters and hangar reproductions were constructed out of rough-cut pine, delivered from the lumberyard across the sound in Manteo. Over the years, the buildings have been rebuilt five times. The last reconstructions of the buildings took place in 1993 in preparation for the 90th anniversary celebration.

For the next seven years, until the visitor center was built in 1960, the Wright brothers' living quarters and hangar would be the focal point of the visitor's experience at the memorial. Visitors could drive up to the site of the First Flight Boulder, park, and tour through the two buildings.

During the 60th anniversary, on December 17, 1963, the First Flight Markers were put in place. But the main event of the 60th anniversary was the long-awaited ribbon cutting and dedication of the 3,000-foot First Flight Airstrip.

The First Flight Boulder that marked the spot where the Wright brothers' flyer took-off on December 17, 1903, was first placed on a small mount with a circle driveway. Visitors to the park could drive around the boulder park and tour the site.

Up until the late 1950s, visitors could drive and park near the First Flight Boulder. Note in this photograph that the boulder faced the monument. In 1966, six years after the visitor center opened, the boulder was turned 90 degrees and now faces the reconstructed buildings.

Before the visitor center opened in 1960, visitors used to convene at the reconstructed hangar and living quarters, located near the First Flight Boulder.

During the 1950s and 1960s, all branches of the military played a major role in the anniversary celebrations. The US Air Force and US Navy were major players in the planning and coordinating of aerial flybys during the ceremonies.

On December 17, 1958, the 55th anniversary of the first flight, the Marine Corps Honor Guard from the Cherry Point Marine Air Station participated in the wreath-laying ceremony at the monument, a tradition dating back to the late 1920s.

In 1958, when this photograph was taken of the monument, the first flight grounds were becoming the focal point of all activities at the park. The new entrance to the park had just been relocated off US Route 158, and more and more people started to visit the national park.

The loop road around the base of the 26-acre hill is almost one mile, making it a good place to walk, jog, and bike. The five parking lots at the base of the hill allowed visitors to drive closer before making the trek up the hill to the monument.

Seven

THE VISITOR CENTER IN 1960

In 1955, the National Park Service developed a 10-year plan to upgrade the national parks. The plan was called Mission 66 and would be completed by 1966, the 50th anniversary of the National Park Service. The Wright Brothers National Memorial Visitor Center is the product of Mission 66. The visitor center was one of its first commissions, and it brought the architectural firm of Mitchell/Giurgola, of Philadelphia Pennsylvania, to national recognition as the first example of the National Park Service's visitor center concept that became standard across America. This photograph was taken on October 29, 1962, by Alexandre Georges.

Bids for the construction of the new entrance off US Route 158 started in early 1958. The contract was awarded to the lowest bidder, Dickerson, a contractor out of Monroe, North Carolina, for the new east park entrance and parking lot for $73,930. The project was completed in December 1958.

In February 1959, the contract for the construction of the visitor center was awarded to the Hunt Contracting Company of Norfolk, Virginia, for a low bid of $257,203.

The most striking feature of the Wright Brothers National Memorial Visitor Center is its dome. Supported by columns, the hollow upper half of the sphere structure encloses the large spacious Flight Room, which houses the reproduction of the Wrights' 1903 flyer.

As early as 1941, it was believed by many involved in the planning of the visitor center that the aviation industry would assist financially in its construction, but none stepped up. Today, the multibillion-dollar industry created by the Wright brothers' achievement at Kitty Hawk shows very little interest in rewarding the origins of the profession and its glorious profitable past.

The building of the Wright Brothers National Memorial Visitor Center was the most dramatic change made to the site. The visitor center resembles an airport terminal and houses original artifacts physically associated with the Wright brothers and their activities at Kitty Hawk.

Two years after the visitor center opened to the public, visitation to the site dramatically increased and would only continue to rise in the next two decades. It became apparent that the building was too little to accommodate the number of visitors to the site.

By the early 1970s, plans were being developed to add an addition to the Wright Brothers National Memorial Visitor Center. This topic has been talked about for the past 40 years with no action taken.

With the approaching 100th anniversary of the first flight in 2003, it was hoped that interest once again would ignite the plans to build a new visitor center museum at the site. Years before the celebration, the First Flight Foundation provided funds to hire an architectural firm for planning and design of the new building, but as fate would have it, the preservation of the Mission 66 structure brought an end to the project.

In January 2001, the visitor center was designated as a National Historic Landmark. The secretary of the interior designated the building because of its exceptional significance as one of the first Mission 66 visitor centers in the National Park Service.

The original parking lot for the visitor center was completed in December 1958. In the early 1980s, additional parking spaces were constructed southward of the existing lot to accommodate the increasing number of visitors to the park.

The exhibits inside the visitor center were also funded, designed, and installed by the Mission 66 initiative. Seen here are two small children viewing the photograph of the first flight at the entrance of the exhibit hall.

The 1960 exhibits inside the visitor center told the story of the Wright brothers from their childhood to the first flights at Kitty Hawk in December 1903. Intermixed in the displays were original Wright brothers' artifacts acquired from the Wright family.

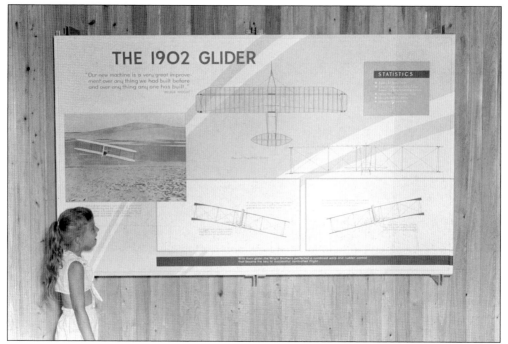

The exhibits inside the visitor center remained in place until the end of the summer of 2016. During the renovation of the visitor center in 2017–2018, the exhibits were updated and replaced. In 56 years, the exhibit was enjoyed by tens of millions of visitors to the park.

As visitation increased over the years, more emphasis was placed on interpreting the story of the Wright brothers. The new Mission 66 exhibits were just the thing to spark the interest of the visitors. Seasonal rangers conducted what were then called "spot talks," short impromptu interpretative programs on the museum exhibits.

Here, seasonal ranger Jerry Cahoon is giving a Flight Room talk in front of the reproduction of the 1903 *Wright Flyer*. Cahoon went on to become a legendary varsity football coach of Manteo High School for many years.

Ranger-led Flight Room talks were the most popular program given at the site throughout the year. In one year, over 1,350 programs were given to 117,000 park visitors. The new renovation of the visitor center in 2017–2018 brought an end to one of the most informative and exciting ranger-led programs given at the park.

During the 57th anniversary, Gen. Benjamin Foulois spoke at the dedication of the Wright Brothers National Memorial Visitor Center on December 17, 1960. General Foulois and two other officers were the US military's first pilots and were trained by Wilbur Wright at College Park, Maryland, on October 26, 1909.

The 64th governor of North Carolina, Luther H. Hodges, addressed the crowd on the rear patio during the dedication ceremony of the Wright Brothers National Memorial Visitor Center on December 17, 1960. Every December 17, since the opening of the visitor center in 1960, the first flight anniversary celebration has been held on the rear patio.

Over the years, many national and international dignitaries have visited the Wright Brothers National Memorial. Pictured here is Lt. Gen. Yoshitoshi Tokugawa (1884–1963) of Japan, second from the left, standing at the site of the first flight. On December 10, 1910, flying a French-built Farman II, Tokugawa was the first person to successfully fly an airplane in Japan.

General Tokugawa traveled all the way from Japan to visit the site and to present a gift from the Japanese people. The gift was a Japanese cherry tree that was planted on the hill near the Wright brothers monument. Standing are, from left to right, park superintendent Robert Gibbs and General Tokugawa with the shovel planting the tree.

Visitors to the park are pictured reading the cast aluminum interpretative wayside exhibit near the reproduction of the Wright brothers' hangar. The exhibit quotes what Orville said about the significance of the first flight in the history of the world.

Visitors to the park were allowed to enter the reproduction of the Wright brothers' 1903 living quarters. The Wrights describe it as their five-room house all in one room—kitchen, dining room, living room, washroom, and bedroom up top.

Before the visitor center opened in 1960, the monument was the point of contact for visitors to the memorial. Ranger-led interpretive programs took place in the base of the monument. From 1983 to 1993, the monument was opened each summer for visitors to climb up to the observation platform on top.

Once the entrance to the park was relocated off US Route 158 in 1958, the loop road around the monument became a one-way road, and the original entrance off Colington was closed. The driveway that once led to the First Flight Boulder was closed, and today, it is a walkway that leads up to the monument.

Stormwater runoff has always been a problem at the site. Even today, during periods of long and heavy rain, large water puddles gather in the open fields. The large puddles linger for days, drawing seabirds to the site to wade in the water. Some of the rangers fondly call the puddles "Lake Wright."

For many years, the park kept the hill well-manicured. During the summer months, maintenance workers would mow the 26-acre hill. Seen here is maintenance worker Willie Simmons mowing the hillside. Simmons went on to become the first African American appointed commissioner for the town of Manteo, North Carolina.

As early as 1934, the original goal of the Kill Devil Hill Memorial Association, which became the First Flight Society in 1966, was to orchestrate the construction of an airport at the site. In 1961, serious discussions and planning for the First Flight Airstrip were beginning to take hold.

By 1930, the National Park Service had received cash donations totaling over $1 million for the purchase of land. But it resisted the airport landing strip and wanted to maintain and preserve the natural scenery and historic resources for the enjoyment of future generations.

Newspaper editor W.O. Saunders, the first president of the Kill Devil Hill Memorial Association wrote, "Wright Memorial . . . without a landing field . . . much like a ship stranded on a reef." In 1961, the State of North Carolina, the Federal Aviation Administration, and the National Park Service funded $133,000 to begin construction of the airstrip. On December 17, 1963, the 60th anniversary of the first flight, it was dedicated and opened for air traffic.

For over 60 years, the Elizabeth City High School Band has performed during the first flight anniversary celebrations. The tradition continues to this day. Band members and drum majorettes still brave the cold winds that brought the Wright brothers to Kitty Hawk at the turn of the 20th century.

The first 1903 *Wright Flyer* was delivered to the park in 1963. The reproduction was built by the American Institute of Aeronautics and Astronautics (AIAA), Washington, DC, chapter. The model took 50 skilled craftsmen three years to build at a cost of over $500,000.

Flight Room talks were a tradition at the Wright Brothers National Memorial. Held in front of the reproduction of the 1903 flyer, located in the Flight Room Auditorium, the talks became very popular. Generations of families brought their children and grandchildren to listen to the presentations.

At the moment Neil Armstrong stepped foot on the moon, legendary Outer Banks photographer Aycock Brown snapped this photograph of the moon over the Wright Memorial. Millions of people throughout the world watched the moon landing on television. (Courtesy of Aycock Brown, July 20, 1969, Outer Banks History Center.)

Like moths attracted to a flame, hundreds of visitors gathered at the Wright Brothers National Memorial Visitor Center to watch man walk on the moon on black-and-white televisions. Little did they know that a piece of the original cloth from the Wrights' 1903 flyer was taken to the surface of the moon that night. (Courtesy of Aycock Brown, July 20, 1969, Outer Banks History Center.)

On July 20, 1969, the night man walked on the moon, hundreds of park visitors packed the Flight Room Auditorium anticipating the historic moment. Imagine the roar of the crowd as they witnessed Neil Armstrong step on the moon and speak the following immortal words: "That's one small step for a man, one giant leap for mankind." (Courtesy of Aycock Brown, July 20, 1969, Outer Banks History Center.)

Tiki torches illuminate and decorate the First Flight Boulder on the night man first walked on the moon. It is hard to believe that only 66 years after the Wright brothers first powered flight at Kitty Hawk, Neil Armstrong walked on the moon. (Courtesy of Aycock Brown, July 20, 1969, Outer Banks History Center.)

In an effort to control growth around the park, the Avalon and Old Dominion Foundation donated $83,000 to the National Park Service for the acquisition of land to extend the park's boundary. Funding from the state was also acquired. An 800-by-2,000-square-foot parcel of land between the monument and US Route 158 was purchased to expand the park boundary to the east.

In 1959, the park enhanced the new entrance off US Route 158 to the east of the park by acquiring 15 acres of land between today's US Route 158 Bypass and the beach road. Over the years, many ideas have come up for the use of this land, but the park is still inclined to keep it in its natural state.

Eight

THE PARK TODAY

For the past 92 years, the Wright Brothers National Memorial has preserved the history of the origins of the world's greatest aviation heritage—a heritage that has inspired generations to achieve the impossible. At the time the park was established by Congress in 1927, very few could have imagined the impact this humble event would have on future world events of the 20th century or the wonder of flight today. Some of the greatest milestones in humankind have occurred since that time—discoveries that have changed the world for better and for the worst. The next 92 years will make the past look like slow motion.

The site has changed completely since the Wright brothers first came in September 1900. Roads, bridges, and the hundreds of thousands of visitors to the Outer Banks each year would now draw them here to conduct their experiments.

Orville's last visit to the site was in April 1939 when he drove down from Ohio in his customized Hudson and picked up his old friend William Tate in Coinjock, North Carolina. They were shown around by Supt. Horace Dough and seemed to enjoy their visit greatly. Orville commented, "Except for the prolific scrub growth . . . I don't see that it has changed much since we were first here. In earlier years there was just sand!"

The park began collecting an entrance fee in February 1987. In the summer months, during high park visitation, fees were collected at the front gate kiosk as cars exited the park, and in the winter months, fees were collected at the information desk as visitors entered the visitor center.

The entrance fee to the park in 1987 was $2 per person with children 15 and under free. Over the years, entrance fees to the park have increased significantly.

From December 11 to December 17, 1978, the park held a seven-day 75th anniversary celebration. Two of the world's legendary aviation historians, Paul E. Garber and Charles Gibbs-Smith, were on the program's schedule and lectured on the Wright brothers' legacy and their influence on the first 75 years of flight.

As part of a tradition that started in 1928, descendants of John T. Daniels, one of the witnesses to the Wright brothers' first flight, place a wreath at the site of the First Flight Boulder. Seen in this photograph are, from left to right, John Wesley Daniels, Hal Daniels, and park employees Peggy Snead and Bebe Burrus Woody.

During the 75th anniversary, two C-141 Starlifters participated in the flyby honoring the Wright brothers. The C-141's first flight was on December 17, 1963, and it remained in service for over 40 years. In 2006, the Starlifter was retired by the US Air Force and was replaced by the C-17 Globemaster.

The highlight of the 75th anniversary was the reenactment of the first flight on December 17, 1978. Ken Kellett, of the Quest for Flight Society, constructed a reproduction of the 1903 flyer from plans he acquired from the National Air and Space Museum Drawings Department. Over 10,000 people witnessed his attempt to fly that morning.

The rededication of the Wright Brothers National Memorial on May 2, 1998, kicked off the centennial celebration. The honorary chairman of the First Flight Commission, Pres. George H.W. Bush, was the guest speaker of the event.

Nearly 20,000 people attended the rededication of the monument to catch a glimpse of Pres. George H.W. Bush. Bush was the first combat naval aviator to become president of the United States.

In honor of Pres. George H.W. Bush, the Navy Band performed during the rededication of the monument. The highlight of the event that evening was the relighting of the rotating beacon on the top of the monument.

For over 58 years since the Wright Brothers National Memorial Visitor Center opened to the public, the back plaza of the visitor center has been the site of numerous special events. Each year, the formal ceremonies of the annual first flight celebration takes place at this site.

In preparation of the centennial celebration, a new brick entrance to the Wright Brothers National Memorial was constructed. The new entrance sign makes a good photo opportunity for park visitors to the site.

The new entrance road was reconfigured to allow the collection of the park entrance fees as cars enter the park. The new fee booths, built in 2001, allowed for a more efficient operation.

The exterior of the Wright Brothers National Memorial has not changed since its construction in 1960. The visitor center has been refurbished twice in the past 20 years. The flat roof design has created many leakage problems inside the building over the years.

The Soaring Society of America's bronze plaque, designed by sculptor Ralph S. Barnaby, was unveiled during the 60th anniversary on December 17, 1963. Barnaby held US Glider Pilot's Certificate No. 1, dated May 26, 1930, and it was signed by his friend Orville Wright. In August 1929, Barnaby broke Orville's American soaring record (9 minutes, 45 seconds in October 1911) by soaring 15 minutes, 6 seconds from Truro, Cape Cod, Massachusetts.

The reconstructed camp buildings were first assembled in 1953 for the 50th anniversary. The structures were rebuilt in 1963 and 1977 because of the harsh beach environment and insect damage. The hangar was rebuilt after it blew down in a storm in 1983, and once again, both structures were rebuilt in 1993 to what is seen today.

The plaque on the First Flight Boulder once faced the monument on the hill. In 1966, the boulder was turned 90 degrees to attract visitors as they approached the site of the first flight.

The First Flight Markers depicting the distance and time in the air of the initial four flights were originally erected in 1963. All four markers were replaced in 1993 in time for the 90th anniversary celebration.

The walkway leading from the boulder to the base of the Big Kill Devil Hill was paved in 2003. In the early days of the monument's construction, there was a road coming from the loop road at the base of the hill, and visitors could drive up to and around the First Flight Boulder.

On the lee, or downwind, side of the Big Kill Devil Hill, the south slope is very steep. Because of the hill's migration toward the southwest, the Big Kill Devil Hill moved 450 feet in 25 years (1903–1928).

Big Kill Devil Hill was the site of nearly 1,000 glider flights made by the Wright brothers between 1900 and 1903. The hill, a live sand dune at the time, has changed significantly. Most of the changes to the site started in the early 1930s in an effort to bring national attention to the Wright brothers' achievements.

This aerial photograph clearly shows how park management kept the site well mowed during the 1980s and 1990s. The well-manicured lawns and hills stand out in this photograph. To the left of the First Flight Airstrip is the only other sand dune, West Hill, used by the Wright brothers during their 1901–1903 glider experiments, that remains at the site today.

The old entrance to the park is still visible today on the south boundary of the park along Colington Road. Just south of Colington Road are more developed areas. The new First Flight High School is visible off the hill to the south of the park.

The First Flight Airstrip averages over 100 airplanes a month during the year. Many new pilots doing their cross-country training fly into the strip to document in their logbooks their visit to the site of the first flight.

The new pilot lounge was built just before the centennial celebration on December 17, 2003. The new lounge has bathrooms and a computer room so visiting pilots can file flight plans and check the weather. Aircraft Owner and Pilot Association (AOPA) donated $300,000 for the construction of the facility.

The First Flight Sculpture was a gift from the State of North Carolina to commemorate the 100th anniversary of the first flight. The full-scale model of the 1903 flyer is a jungle gym for children to climb on and play on.

The *First Flight* sculpture, located on the south side of Big Kill Devil Hill, depicts the scene of the first flight on the morning of December 17, 1903. The life-size statues of the four men and the 16-year-old boy who witnessed the flights that morning were created by sculptor Stephen H. Smith, who also created the statue of Richard Etheridge, the first African American keeper in the US Lifesaving Service, located in Manteo, North Carolina.

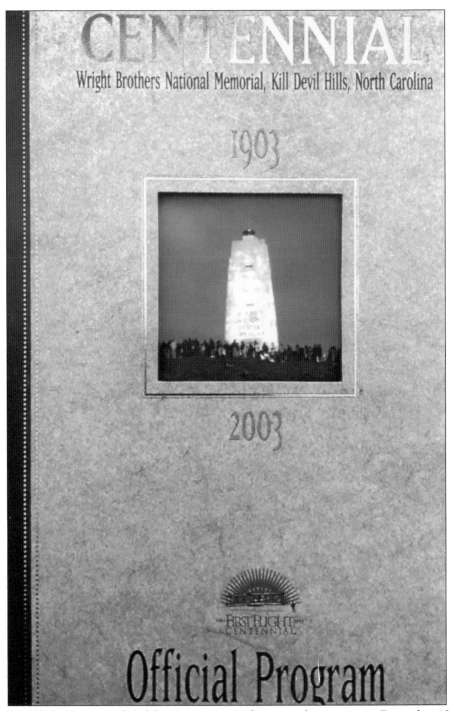

CENTENNIAL

Wright Brothers National Memorial, Kill Devil Hills, North Carolina

1903

2003

FIRST FLIGHT CENTENNIAL

Official Program

The First Flight Centennial Celebration was a six-day event, beginning on December 12 and ending on December 17, 2003. To inspire the next generation of aviators, the theme of the first day was "Igniting the Imagination." The theme of the last day of the celebration was "Twelve Seconds that Changed the World," which honors the first flight of the Wright brothers. The centennial celebration culminated with a reenactment of the first flight.

This is an aerial view of the Wright Brothers National Memorial during the centennial celebration in 2003. The First Flight Centennial Pavilion Exhibit Hall, an 8,000-square-foot building, one of the three structures that comprised the $2 million, 20,000-square-foot pavilion, housed exhibits sponsored by NASA, the Federal Aviation Administration, and the Experimental Aircraft Association (EAA). For example, the EAA's Countdown to Kitty Hawk was a popular exhibit.

Visitors from around the world attended the centennial celebration. The once-in-a-lifetime event brought aviation pioneers and legends from all aspect of aviation and aerospace backgrounds. Neil Armstrong, first man to walk on the moon; Chuck Yeager, first man to break the sound barrier; pilot Robert Morgan, captain of the *Memphis Bell* (the first B-17 heavy bomber to complete 25 combat missions over Europe in World War II); and 100 aviation pioneers were honored during an event in front of over 40,000 people on December 15.

The main stage of the First Flight Centennial Celebration was where all the formal ceremonies and official activity took place during the event. Over 200,000 people from all around the world attended the six-day celebration.

One of the highlights of the event was the reenactment of the first flight at 10:35 a.m., the documented time the Wrights made their first flight on December 17, 1903. On that day, it rained, the temperatures warmed up, and the wind died out, so the flyer could not get off the ground.

Although the flyer did not get off the ground that day, it did fly two times other times. The first successful flight was the Thursday before Thanksgiving 2003, and it flew about 119 feet. On the second flight in early December 2003, a lady pilot who was in the running to fly it the day of the centennial, December 17, 2003, crashed without being injured.

The US Lifesaving Service, predecessor to the US Coast Guard, played an important role in the Wright brothers' story at Kitty Hawk. Three members witnessed the first flight on December 17, 1903, and one member, John T. Daniels, snapped the photograph of the first flight that morning. Coast Guard aircraft performed a flyby honoring the Wright brothers during the centennial celebration.

Hundreds of aircraft participated in the flyby during the 100th anniversary celebration. Every aircraft in the military's inventory flew over the site on December 16. In this photograph depicting a "Heritage Flight" over the monument, two P-51 Mustangs flew in formation with an F-15 Strike Eagle and an F-16 Falcon.

The Douglas DC-3 was one of the airplanes that participated in the flyby during the centennial celebration. The DC-3 revolutionized air travel in the 1930s. It was operated by all the US airlines of its time—American, United, Trans World Airlines, Delta, and Eastern—setting the standard for the modern American air travel industry.

Pres. George W. Bush attended the celebration on December 17, 2003. Air Force One landed at the Coast Guard Air Base Station in Elizabeth City, North Carolina, and Marine One flew the president and his entourage to the park.

The V-22 Osprey made its first appearance at the centennial celebration in December 2003. The tilt-rotor aircraft has both vertical takeoff and landing (VTOL) and short takeoff and land (STOL) operational capabilities.

Every morning as part of the opening ceremony, a parachutist would jump from a plane with the giant American flag suspended beneath him as the national anthem played.

THE EVENT OF THE CENTURY

FIRST FLIGHT
CENTENNIAL
THE WRIGHT BROTHERS FIRST FLIGHT
DECEMBER 17, 1903 • DECEMBER 17, 2003
KITTY HAWK, NORTH CAROLINA

The First Flight Centennial Celebration was promoted as "the event of the century." This poster, printed to market the event, shows a young Wilbur Wright (1867–1912) and an old Orville Wright (1871–1948). Orville outlived his brother Wilbur by 36 years.

After a two-year renovation, the Wright Brothers National Memorial Visitor Center reopened in the fall of 2018. The $5.8 million rehab project renovated everything from the roof to the mechanical and electrical systems of the 58-year-old building.

The new visitor center information desk has been relocated so that the ranger manning the desk can greet visitors as they enter the building. To provide easy access to park information and orientation, part of the desk has been lowered to accommodate visitors in wheelchairs.

The original exhibits inside the visitor center were installed in 1960 when the building first opened to the public. The park spent an additional $1.5 million to replace and upgrade the exhibits.

The majority of the visitors to the site come in family units. The new exhibits tell the story of the Wright brothers' close connection with their family members and how their interaction played a pivotal role in their ultimate success.

Wilbur was working late one night in the bicycle shop when a customer dropped in to buy an inner tube for a tire. While playing with the box and talking to the customer, Wilbur discovered "wing warping." In the exhibit of the Wrights' bicycle shop, visitors can reenact that moment when twisting the inner tube box that inspired the idea of wing warping, the forerunner to the modern aileron.

After three years of flying gliders off the sand dune at the site, the Wright brothers solved the problem of human flight in their 1902 glider. The 1902 exhibit tells the story of hard work, dedication, and perseverance that led to their success on the morning of December 17, 1903.

Orville wrote, "As children we were always encouraged to pursue intellectual interest . . . to investigate whatever aroused our curiosity." The new exhibits are designed to inspire the next generation of children to nurture big dreams and make them come true, just like the Wright brothers did over 100 years ago.

In the new exhibits, questions that are designed to make visitors think about what it takes to be successful and to achieve goals in life are presented to them. Here, the visitors are asked the following question: How do you make your dreams a reality or who inspires you?

Wind, sand, and a dream of flight brought the Wright brothers to Kitty Hawk, but it was the people who lived here that kept them coming back. The sewing machine display tells the story of the bond between the brothers and the people of Kitty Hawk. The locals' hospitality helped form lifelong friendships with the Wrights.

Locals told the story of seeing the Wright brothers running down the beach with their arms spread wide imitating a bird in flight. In this exhibit at the rear of the hall, children can experience the motion of a soaring bird's wings in flight.

In 4.5 years, the Wright brothers solved the problems of flight that had alluded humankind for thousands of years. In this exhibit, visitors see the evolution of the machines from 1899 to 1902.

Every December 17, during the annual first flight celebration, the First Flight Society inducts into the Paul E. Garber First Flight Shrine a person or persons who have achieved a significant "first" in aviation/aerospace. This exhibit is updated every year to display the portrait of last year's inductee.

The building was renovated and restored to its original 1960s look. All efforts were made to preserve as much of its original fabric as possible. The cypress panels, curved walls, and exterior concrete artwork at the entrance of the building were all kept intact.

With the use of new technology, this state-of-the-art 16-screen panel wall can change to highlight different aspects of the Wright brothers' work. This stunning visual platform will enhance the visitors experience for years to come.

Inside the Flight Room, the reproduction of the 1903 flyer is awe-inspiring. Commissioned by Harry Combs, the full-scale flying model was built in 2003 by the Wright Experience, in Warrenton, Virginia. Combs, CEO of Gates Lear Jet, Inc., and author, spent $1.2 million and donated the flyer to the National Park Service on December 17, 2003.

The model of the 1903 flyer on display in the Flight Room Auditorium was built by the Wright Experience. Working under contract with the Discovery of Flight Foundation and using reverse engineering, a team of experts examined original parts, photographs, and artifacts to create the most accurate flying model of the *Wright Flyer* ever built.

The interpretive programs at the site are now conducted outside on the back patio of the visitor center, weather permitting. Before the restoration of the visitor center, ranger programs were conducted inside the building in front of the 1903 reproduction.

The Wright Brothers National Memorial has been called a mecca for aviation enthusiasts from around the world. Every year, hundreds of thousands of people from every walk of life visit the site to pay homage to two brothers that changed the world.

Legendary historian Darrell Collins served almost 39 years at the Wright Brothers National Memorial and retired in January 2017. He is now the owner/operator of company called A Legacy of Greatness (ALOG) and still lectures around the country on the Wright brothers and early aviation.

Doug Stover served as cultural resource manager and historian at Cape Hatteras National Seashore, Wright Brothers National Memorial, and Fort Raleigh National Historic Site and is now retired from the National Park Service. He once managed the historic resources that included 68,746 Wright brothers artifacts. Here, Stover is holding one of the most innovative aspects of the 1903 flyer, its propellers. On display inside the visitor center is one of the original propellers from the *Wright Flyer*. The propellers off the 1903 flyer were used on the next airplane the brothers built in 1904. The propeller was damaged during aborted takeoffs at Huffman Prairie in Dayton, Ohio.

Neil Armstrong presents park superintendent Larry Belli with a piece of cloth from the wing of the Wright brothers 1903 flyer and a small piece of Spruce from the propeller that were taken to the moon in 1969. This artifact came full circle when it was donated to the Wright Brothers National Memorial on December 16, 2003, during the centennial celebration. Today, displayed at the visitor center in a small box to protect it from UV light damage, visitors can open the exhibit to see the cloth and a small piece of the propeller.

The original engine crankcase of the 1903 flyer was damaged after the end of the fourth flight on December 17, 1903. In a metal analysis of the engine crankcase, it showed that it exhibited precipitation hardening of aluminum alloy, which is a heat treatment process that increases the strength of aluminum. The crankcase was made-up of 80 percent aluminum and 20 percent copper.

The Wright brothers started their bicycle business in 1892. At first, they bought and sold bicycles but later designed and manufactured them as well. With the type of machinery they had in their shop—lathes, drill press, and hand tools—the Wrights could manufacture bicycle parts, like the hub pictured above.

In the early 1900s, sprockets and chain drives were standard components used to transfer the power produced by the engine to turn the propellers and move the flyer forward through the air.

In 1913, Orville conducted wind tunnel tests on many shapes as well as wing shapes. These two sphere-shapes objects were tested to determine their aerodynamic properties.

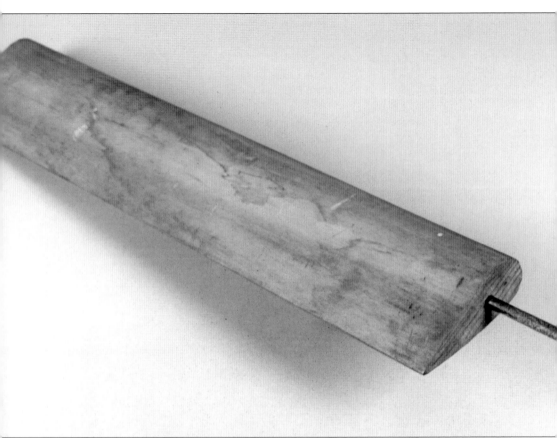

On display at the Wright Brothers National Memorial Visitor Center is an original wooden airfoil used by Orville Wright in his wind tunnel experiments in 1913. The airfoils used in their earlier wing tunnel experiment in the winter of 1901 were made of metal and are now in the collection of the Franklin Institute in Philadelphia.

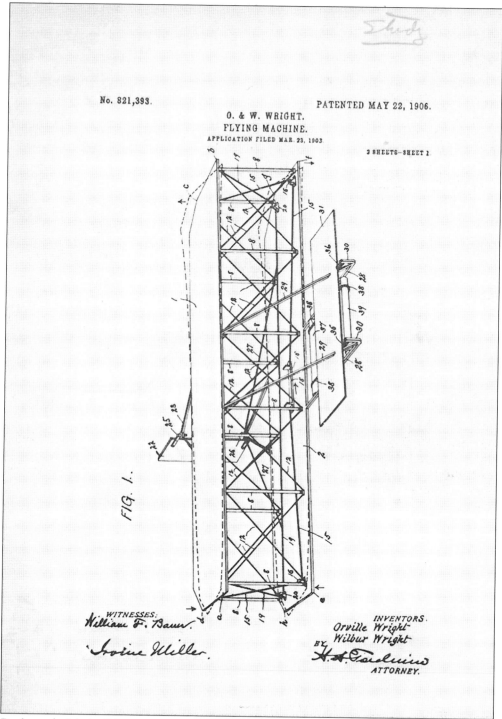

No. 821,393.

PATENTED MAY 22, 1906.

O. & W. WRIGHT.
FLYING MACHINE.
APPLICATION FILED MAR. 23, 1903.

3 SHEETS—SHEET 1.

FIG. 1.

WITNESSES:
William F. Bauer.
Irvine Miller.

INVENTORS.
Orville Wright
Wilbur Wright
BY H. A. Toulmin
ATTORNEY.

By the end of their gliding season at Kitty Hawk in 1902, the Wright brothers had already solved the major problems (lift and control) of mechanical flight. They applied for their patent in March 1903 based on the construction and mechanisms of control of their 1902 glider, not the 1903 flyer. US Patent No. 821,393 was granted on May 22, 1906.

121

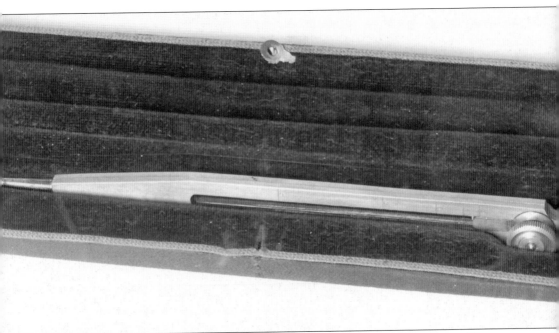

This drafting tool, a compass, was used by Orville Wright to draft technical drawings. The compass can also be used to make circles and arcs. As a divider, it can be used to measure distance on a map, for mathematics, and for navigation.

Pictured above is a stainless steel map depicting the history of aviation from 1903 through 1928. This map of the world was an original feature inside the Wright Brothers National Memorial. Located on the second landing of the monument, it depicts aviation milestones during the first 25 years of powered fight. The list begins with the first flights at Kitty Hawk in 1903 and ends with the first transpacific flight from Oakland, California, to Brisbane, Australia, by Charles Kingsford Smith.

This small piece of the strut from the 1903 flyer is autographed as follows: "Compliments of Orville Wright, Wilbur Wright." It is one of the park's most prized artifacts. Years ago, when the visitor center first opened, a similar piece on display was stolen. This is one of many artifacts that is not on display to the public.

Titled "1903 Kitty Hawk to Mars 1978," these stamps are part of the museum collection and were created to honor the 75th anniversary of the first flight. The stamps commemorate man's aerospace achievements over the past 75 years of flight.

Every year in July, the Wright Kite Festival takes place at the Wright Brothers National Memorial. The two-day kite festival is sponsored by the National Park Service in partnership with Kitty Hawk Kites.

Wright Brothers National Memorial

Wind, sand, and a dream of flight brought Wilbur and Orville Wright to Kitty Hawk, North Carolina. Here, after four years of experimentation, they achieved the first successful controlled airplane flights on December 17, 1903. With courage and perseverance, relying on teamwork and the scientific process, what they achieved changed our world forever.

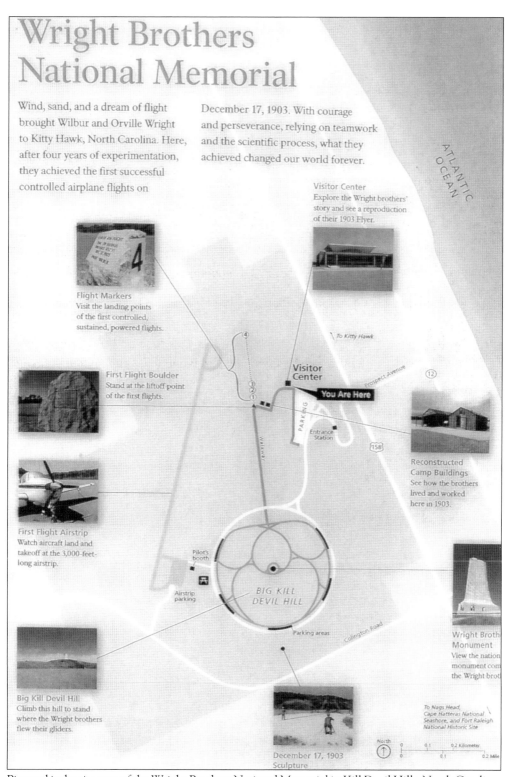

Visitor Center
Explore the Wright brothers' story and see a reproduction of their 1903 Flyer.

Flight Markers
Visit the landing points of the first controlled, sustained, powered flights.

First Flight Boulder
Stand at the liftoff point of the first flights.

First Flight Airstrip
Watch aircraft land and takeoff at the 3,000-feet-long airstrip.

Big Kill Devil Hill
Climb this hill to stand where the Wright brothers flew their gliders.

Reconstructed Camp Buildings
See how the brothers lived and worked here in 1903.

Wright Broth Monument
View the nation monument com the Wright brot

ATLANTIC OCEAN

To Kitty Hawk

Visitor Center

You Are Here

Prospect Avenue

Parking

Entrance Station

158

12

EASTERN

Pilot's booth

Airstrip parking

BIG KILL DEVIL HILL

Parking areas

Colington Road

December 17, 1903 Sculpture

To Nags Head, Cape Hatteras National Seashore, and Fort Raleigh National Historic Site

North

0   0.1   0.2 Kilometer
0   0.1   0.2 Mile

Pictured is the site map of the Wright Brothers National Memorial in Kill Devil Hills, North Carolina.

Discover Thousands of Local History Books
Featuring Millions of Vintage Images

Arcadia Publishing, the leading local history publisher in the United States, is committed to making history accessible and meaningful through publishing books that celebrate and preserve the heritage of America's people and places.

Find more books like this at
www.arcadiapublishing.com

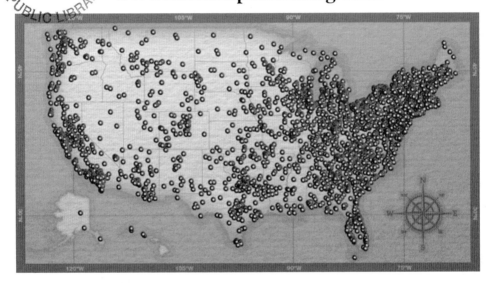

Search for your hometown history, your old stomping grounds, and even your favorite sports team.

Consistent with our mission to preserve history on a local level, this book was printed in South Carolina on American-made paper and manufactured entirely in the United States. Products carrying the accredited Forest Stewardship Council (FSC) label are printed on 100 percent FSC-certified paper.

MADE IN THE